Olena Tokmylenko

Planning Bicycle Infrastructure By Quickest Or Easiest Route Method

Olena Tokmylenko

Planning Bicycle Infrastructure By Quickest Or Easiest Route Method

LAP LAMBERT Academic Publishing

Impressum / Imprint

Bibliografische Information der Deutschen Nationalbibliothek: Die Deutsche Nationalbibliothek verzeichnet diese Publikation in der Deutschen Nationalbibliografie; detaillierte bibliografische Daten sind im Internet über http://dnb.d-nb.de abrufbar.
Alle in diesem Buch genannten Marken und Produktnamen unterliegen warenzeichen-, marken- oder patentrechtlichem Schutz bzw. sind Warenzeichen oder eingetragene Warenzeichen der jeweiligen Inhaber. Die Wiedergabe von Marken, Produktnamen, Gebrauchsnamen, Handelsnamen, Warenbezeichnungen u.s.w. in diesem Werk berechtigt auch ohne besondere Kennzeichnung nicht zu der Annahme, dass solche Namen im Sinne der Warenzeichen- und Markenschutzgesetzgebung als frei zu betrachten wären und daher von jedermann benutzt werden dürften.

Bibliographic information published by the Deutsche Nationalbibliothek: The Deutsche Nationalbibliothek lists this publication in the Deutsche Nationalbibliografie; detailed bibliographic data are available in the Internet at http://dnb.d-nb.de.
Any brand names and product names mentioned in this book are subject to trademark, brand or patent protection and are trademarks or registered trademarks of their respective holders. The use of brand names, product names, common names, trade names, product descriptions etc. even without a particular marking in this works is in no way to be construed to mean that such names may be regarded as unrestricted in respect of trademark and brand protection legislation and could thus be used by anyone.

Coverbild / Cover image: www.ingimage.com

Verlag / Publisher:
LAP LAMBERT Academic Publishing
ist ein Imprint der / is a trademark of
OmniScriptum GmbH & Co. KG
Heinrich-Böcking-Str. 6-8, 66121 Saarbrücken, Deutschland / Germany
Email: info@lap-publishing.com

Herstellung: siehe letzte Seite /
Printed at: see last page
ISBN: 978-3-659-46828-5

DEDICATION

I dedicate this work to my parents, Sergey and Tatyana Tokmylenko, who always believe in me and give an extraordinary support in everything I do.

ACKNOWLEDGEMENTS

I would like to thank my academic advisers, Dr. Eric Morris and Dr. Barry Nocks, for willingness to work with me throughout this project, and especially Professor Stephen Sperry for valuable guidelines and support in time of confusion. I also wish to thank Dmytro Konobrytski and Viktor Zagreba for their essential inputs into development and validation of my model.

I would like to thank Fulbright exchange program for giving me an opportunity to study in the US.

I want to express my love and gratitude to my family and friends for their belief and encouragement throughout this program. A special thanks to Aleksandr Chernyshov without whom I would never be strong enough to complete this program.

TABLE OF CONTENTS

Page

DEDICTATION..3

ACKNOWLEDGEMENTS 5

LIST OF TABLES 8

LIST OF FIGURES..9

CHAPTER

1. INTRODUCTION...8

2. CURRENT STATE OF RESEARCH..10

    INTRODUCTION 10

    CURRENT STATE OF BICYCLE INFRASTRUCTURE PLANNING 10

    TYPE OF CYCLISTS 11

    FACTORS THAT AFFECT CYCLING 14

    TRAVEL TIME AND TRAVEL SPEED 16

    ENERGY EXPENDITURE AND BICYCLING POWER 18

    CONCLUSION..19

3. RESEARCH QUIESTION AND OBJECTIVES 21

4. METHODOLOGY...22

5. DATA AND RESEARCH SCENARIOS 28

6. DISCUSSION OF THE RESULTS 33

    SELECTION OF QUICKEST ROUTE...33

    VALIDATION OF THE RESULTS 38

    METHODS OF INFRASTRUCTURE PRIORITIZATION 40

7. CONCLUSION 44

    APPLICATION TO PRACTICE 44

    LIMITATIONS OF THE RESEARCH 45

    NEXT STEPS 45

REFERENCES..46

1. INTRODUCTION

In 1990 the Federal Highway Administration set a goal of doubling the share of pedestrian and bike trips (from 7.9% to 15.8% of all trips), simultaneously reducing the number of fatalities and injuries among bicyclists and pedestrians by 10% (US Department of Transportation, 2012). In 2009 it more than doubled federal funding available for pedestrian and bicycle improvements (figure 1.1). In March 2009 Transportation Secretary Ray LaHood assured bicycle advocates at the National Bike Sumit that he and president Obama "will work toward an America where bikes are recognized to coexist with other modes and to safely share our roads and bridges" (Fried, 2009). A few weeks ago (February 2013) LaHood announced that Federal Highway Administration will develop its own bicycle and pedestrian safety standards for the first time (Nettler, 2013).

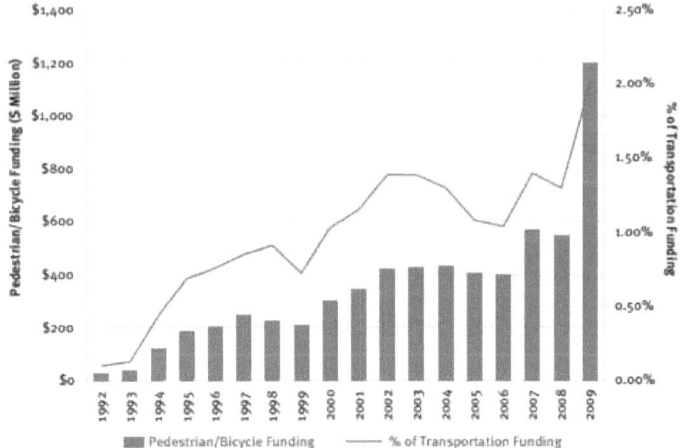

Figure 1.1: Federal Pedestrian and Bicycle Funding, 1992 – 2009 (US Department of Transportation, 2010)

However, even with increased funding, building bicycle infrastructure is a very expensive proposition. Many cities that have adopted bicycle master plans or included bicycling in a city's transportation strategy have used an opportunistic approach to infrastructure development (Litman et al., 2005). This means that they place bicycle ways and appropriate bicycle marking when existing roads are being redesigned. Although this method reduces the cost of facility development and allows using existing funds, it often results into array of randomly distributed strips of bikeways that are not connected into coherent network. In this situation, achieving a goal where all roadways can serve as appropriate bicycle facilities may take several decades. Cities that are constrained by limited road funds, meaning all cities, need a tool to prioritize roads and streets for bicycle infrastructure development. The question that arises is what criteria should be used to define priorities for infrastructure locations. When we are talking about bicycle parking then it is definitely needs to be located at key destinations and thus bicycle ways should connect those destinations one to another (American Association of State Highway and Transportation Officials, 2012). But the most important criterion for selecting a route should be based on riders' preferences to maximize the use of the bicycle network.

Based on careful literature review that follows, this research defines travel time as the most important factor for the utilitarian cyclist or those who use bicycles for a purpose other than simply to enjoy the ride. The question that arises in this research is how to select a bicycle route based on the minimum travel time of bicycling trip. To answer this question I have studied what affects bicycling travel time the most, and learned that a rider's speed depends heavily on physical power of the rider and road topography. I then developed a model that allows predicting changes in travel speed based on road slope and a rider's maximum power, and calculated travel time for each road segment. I applied my model to part of Washington, DC's road network and selected quickest routes between five stations of the Capital Bikeshare system, which is Washington's bikesharing system which allows riders to pick up and drop off rental bikes at locations throughout the city. I then combined all routes into one route network and identified some road segments that are more convenient than others. These are the road segments that allow for the quickest way between many destinations and should be given higher priority for bikeway infrastructure development than others. The last chapter of this paper provides recommendations for planners on how to use this method for their bicycle-oriented initiatives and highlights opportunities for further research.

2. CURRENT STATE OF RESEARCH

INTRODUCTION

This chapter takes a look at current tools available to city planners who plan bicycle infrastructure. It then discusses important factors that affect people's decision to ride a bike and determines that there is a gap between planning methods currently in use and riders' concerns. This chapter also pulls together important concepts or terms that are used in this research and provides a theoretical basis for my research methodology.

CURRENT STATE OF BICYCLE INFRASTRUCTURE PLANNING

In the 1970s the Federal Highway Administration carried out research called "Safety and Locational Criteria for Bicycles Facilities". The purpose of the research was to develop recommendations for planning agencies on how to choose locations and designs for bicycle facilities. One of their reports groups criteria that have to be considered while choosing facility location into primary user-related, other user-related, and general. Primary user-related criteria include: potential use, basic lane width, connectivity and directness, safety, grades (i.e. slopes), and physical barriers (Smith Jr, 1975). The study proposes a comprehensive approach for planning bike facilities, from discussing why planners should develop bicycle infrastructure to offering practical recommendations for physical design. This report marked the beginning of bicycle infrastructure planning in the US.

Currently there are three main documents that are used by urban planners when developing bicycle facilities. The first is the *Guide for Development of Bicycle Facilities, 4th Edition* by American Association of State Highway and Transportation Officials (AASHTO) (American Association of State Highway and Transportation Officials, 2012). The guide covers the main elements of the bicycle planning process, emphasizes the importance of safety, and provides some design guidelines for infrastructure development.

The guide suggests that the final goal is to make all roadways accessible and suitable for bicyclists, except those where cycling is prohibited. However, since this goal cannot be achieved immediately, the guide suggests considering the following factors when deciding where to place improvements to bicycle infrastructure: user needs; motor vehicle traffic volumes; vehicle mix (e.g. passenger cars, trucks, heavy trucks, etc.), and speeds; constraints and physical barriers; connections to land uses and access to key destinations; directness of route; logical sense of route; intersections; aesthetics; spacing or density of bikeways; safety;

security; and overall feasibility. The guide offers several technical analysis tools to make better decisions about bicycle infrastructure development.

Data collection and *flow analysis* is a method of collecting statistical data about current bicycle volumes and patterns. These data allows planners to understand the number of riders and patterns of bicycling in their locality, to analyze demographics, and to forecast travel demand. Bicycle infrastructure can be then developed considering knowledge about riders in the area or following their current routes. This method is especially useful for areas with large current volumes; however, areas with a very low level of bicycling may not have enough input data for analysis.

Level of service analysis or *compatibility index* is a tool developed by the Federal Highway Administration to evaluate "the comfort levels of bicyclists on the basis of observed geometric and operational conditions on a variety of roadways" (Harkey, Reinfurt, Knuiman, Stewart, & Sorton, 1998). The index was obtained through empirical study of visual survey responses of respondents who evaluated their comfort level when watching video record of riding a bicycle. Examiners discovered that presence of bike lanes, the width of the shoulder or curb lane, the presence of on-street parking, etc. affect bicycling's compatibility with a road from the human perspective. However, the video survey method that was used does not consider the physical involvement of riders and thus limits factors to only those that can be observed visually by participants.

Safety analysis is a method to plan appropriate bicycle facilities based on crash data analysis to improve the level of safety along major corridors. AASHTO recommends using an Intersection Safety Index, which helps to identify intersections that are more or less dangerous to cyclists and prioritize intersection improvements or decide to reroute bikeways (American Association of State Highway and Transportation Officials, 2012).

Authors of the guide distinguish *GIS-based network planning* as a separate tool for bicycle infrastructure planning; however, GIS systems can be used to store and analyze data for all the methods outlined above. Moreover, GIS systems allow integrating different prioritization tools for a specific network and choosing the most appropriate.

The last tool AASHTO recommends considering for bicycle infrastructure improvement is *cost-benefit analysis*, where cost is determined as one-time construction cost and annual operating costs for a bicycle facility for some period of time, and benefit is determined based on some kind of measured economic benefits (e.g., time savings, increased livability, decreased health costs, a more enjoyable ride, etc.).

Analysis of the AASHTO document has shown that, although the guide provides valuable directions for planners in developing bicycle infrastructure, the methods for infrastructure prioritization are limited and leave a lot of space for planners' creativity. These methods do not cover all factors that have to be considered when developing bikeways, and especially those related to specific locations of bicycle routes, namely constraints and physical barriers; directness of route; logical sense of route; aesthetics, and spacing or density of bikeways.

The second document widely used by planners to develop bicycle infrastructure is the *Manual on Uniform Traffic Control Devices* or MUTCD (U.S. Federal Highway Administration, American Traffic Safety Services Association, American Association of State Highway and Transportation Officials, & Institute of Transportation Engineers, 2010). This is the official document from the Federal Highway Administration that defines standards for signs, signals, and marking for bicycle facilities' traffic control. The document does not provide recommendations on where to locate bicycle facilities but rather regulations on how to integrate such facilities into a road network.

The third document is National Association of City Transportation Officials *(NACTO) Urban Bikeway Design Guidelines* (National Association of City Transportation Officials, 2012). The document provides an array of recommendations for bike lane placement, intersection improvement, signing and marking, and other aspects of bicycle facility development supported with pictures and best case examples from different cities. The document also offers a master reference matrix of research and studies that can help to plan bicycle facilities.

Of course there are other studies that discuss possible ways of developing bicycle facilities (King, 2002; Litman et al., 2005; Pucher, Dill, & Handy, 2010). The methods they all discuss more or less fall into categories distinguished by AASHTO. The main emphasis is on safety and the economic feasibility of the projects.

However, it is hard to believe that safety is the only factor that affects a rider's decision to cycle. Separate lanes or bikeways will allow the cyclist to feel safer; however, will not necessarily encourage him to cycle. There are still many other factors that affect cycling like distance between origin and destination, ease of ride, purpose of ride, etc. To understand what to consider when developing bicycle facilities, I analyzed types of riders and factors that affect people's decision to cycle.

Many agencies use the classification of cyclists proposed by the Federal Highway Administration (FHWA) based on bicycle stress levels, meaning how comfortable a cyclist feels riding on a particular road segment (Harkey, Reinfurt, Knuiman, Stewart, & Sorton, 1998) . According to it, bicyclists can be grouped into three categories:

- Group "A" – advanced or confident cyclists. This group includes adult riders who are confident riding in mixed environments and can coexist with motorized vehicles. These riders pay less attention to facility quality and require minimum safety levels.

- Group "B" – basic cyclists. These are teenagers or young adults who are less confident riders and require higher levels of facility development and safety; they prefer to ride on a separate lane or way.

- Group "C" – children, typically accompanied by parents, the most vulnerable group of riders.

AASHTO evaluates bicyclists by their level of skills and comfort (confident and less confident), as well as by age (children, adults, and senior users) (American Association of State Highway and Transportation Officials, 2012).

Even though an understanding of these characteristics is important when planning bike facilities, these classifications do not take into account the physical abilities (except for age) of the rider for accomplishing rides of different levels of difficulty. While all people who have been riding for a while and can confidently travel in motorized traffic will be considered advanced cyclists, not all of them will be physically ready to undertake a route with steep slopes or other obstacles. There is a need to look more broadly at cyclists' abilities when planning bike facilities.

Another factor that affects bicyclists' behavior is trip purpose. AASHTO and FHWA distinguish utilitarian (nondiscretionary) and recreational (discretionary) trip purposes. Utilitarian cyclists are those who use bicycle as transportation mode to get to their destinations. For this type of user a reasonable combination of distance and travel time is essential. Recreational riders, on the other hand, are those who see pleasure of activity as the main purpose of trip. Smith (1975) notes that utilitarian cyclists tend to maximize efficiency of the trip, while recreational riders value safety and the quality of ride.

The method used in a Vancouver survey (Winters, 2011) segmented the population into four categories based on ridership frequency. Regular cyclists are those who travel by bicycle at least once a week ($\geq$ 52 trips per year), frequent cyclists cycle at least once a month

(12-51 trips per year), occasional cyclists cycle at least once a year (1-11 trips per year), and potential cyclists have not cycled in a year prior to survey, but have access to a bike and would consider bicycling in future. The study shows how the frequency of ridership reflects motivators for and deterrents to cycling. We can assume that regular cyclists are more physically developed, since regular bicycling has proven to have beneficial effect on health and athleticism (Tolley, 2003).

There are also other ways on how to divide riders into groups. For example, some sport-oriented websites (e.g. www.cyclingpowerlab.com) distinguish cyclists by years of active ridership or level of proficiency (e.g. non-racing cyclist, beginner, elite racing cyclist). The categories into which planners sort cyclists should reflect the purposes of the study. However, my investigation has shown that there is no classification that directly characterizes physical abilities of the rider for different difficulty levels in planning documentation. The next step is to analyze factors that affect people's decision to cycle.

FACTORS THAT AFFECT CYCLING

Recently many researchers have concentrated their attention on what encourages or discourages people from bicycling (Broach, Gliebe, & Dill, 2011; Jaffe, 2012; Sener, Eluru, & Bhat, 2009a; Stinson & Bhat, 2003; Winters, 2011) . One the most comprehensive studies on this topic was completed in Vancouver, Canada (Winters, 2011). The purpose of the "Cycling in Cities" survey was to determine the potential motivators of and deterrents to bicycling. Both motivators and deterrents were placed in categories such as: vehicles; lane markings; intersections; distances; hills and connections; road surfaces and maintenance; aesthetics and access; coordination with transit; social interactions; safety; weather and darkness; legislation; and information and incentives. The usefulness of this survey was that it looked fairly at the factors that affect people's decision to bicycle and evaluated the positive and negative elements in each category. The respondents noted that off-street paths are strongly desired because they provide separation from traffic, noise, and air pollution; riders also noted paths should be flat, lit, and provide direct access to the final destinations. The research showed that the ease of cycling was among the top factors that have the strongest potential influence on cycling, together with safety and aesthetics. What seems controversial is that the physical challenge of a trip was registered as having little influence, especially among the most frequent riders. This can be explained by the assumption that people will be encouraged to cycle if the route is easier; however, if they decided to cycle they are not likely to quit because of the physical challenge of the route. Also, regular cyclists develop muscular strength, which can explain why physical difficulty is less important to frequent riders than to others.

Many researchers have considered travel time to be an important factor for bike commuting (Sener, Eluru, & Bhat, 2009a; Sener, Eluru, & Bhat, 2009b; Smith Jr, 1975; Stinson & Bhat, 2003; Winters, 2011). However, Sener (2009) and Smith (1975) argue that, although travel time is highly important for biking, it is relevant to commute-related trips only. Winters (2011) in her research finds that majority of respondents consider 30 minutes to be the optimal time for a bike trip. Sener (2009) notes that based on stated preferences survey in Texas, travel time is more significant for the younger population (18-34), who prefer trips that are shorter in terms of duration than the older population.

Topography is another important factor mentioned by researchers. According to a bike survey from Austin, Texas, among bicyclists commuting to work females tend to avoid hilly routes, while males prefer steep slopes to flat topography and moderate hills to steep slopes. At the same time, women traveling for recreational purposes prefer routes with moderate hills and men significantly prefer steep hills over moderate hills, and moderate slopes over flat terrain (Sener, Eluru, & Bhat, 2009a)   .

Both Smith (1975) and Winters (2011) identify topography as the most important factor determining whether people will ride for commuting purposes. A bike study based on GPS data collection accomplished in Portland, OR (Broach, Gliebe, & Dill, 2011)   has shown that cyclists will rather cycle 1.76 miles of flat route than 1 mile uphill of with a 2-4 percent slope. These results contradict to those found by Sener et al (2009a). However, the study in Texas used a preference survey while the study in Portland analyzed actual travel data collected by GPS. It can be assumed that people may report they prefer steeper topography because of their desire to be fit, however, in actuality they may not necessarily ride on steeper topography. Broach et.al also identified that travel patterns of riders are based on the following grades: flat to 2 percent uphill slopes, 2-4 percent uphill slope, 4-6 percent uphill slope, and more than 6 percent slope. These ranges represent categories of rode slope that affect the likelihood that people will choose one or another segment for their trip, where it is more likely people will ride on a flat terrain, and it is not likely riders will choose the route with slopes of more than six percent. Unfortunately, the study does not explain the difference in behavior based on the length of uphill slopes.

The differences in choices explained above show that travel time and topography are important for bicyclists and are mistakenly ignored by transportation planners when planning bicycle infrastructure.

Bicycling travel time has not been studied well. A 1999 review on state of the art in the field of bicycle science, operation, and design does not mention any single piece of research on determining riders' travel times (Taylor & Davis, 1999). Yet current methods on how to estimate travel time of motorized transport can be analyzed and applied in part to predict bicycle travel time.

Basically, travel time for a motorized vehicle in urban conditions is combination of free flow travel time and time for delay. Free flow travel time is calculated as the distance divided by free flow speed, where free flow speed varies based on driving behavior, speed limits, weather conditions, spacing between intersections, etc. Free flow travel time for motorized vehicles can be estimated based on assumptions about free flow speed (e.g. the speed limit), but a bicycle rider's free flow speed cannot be simply assumed, because it is limited by the maximum possible power that can be applied to operate the vehicle and can vary significantly from rider to rider. Delay is the second part of travel time equation; for nonmotorized vehicles it can be caused by different factors (e.g. parked vehicles, intersection signals, pedestrian crossings, etc.) (Zheng, 2011) and can be similar to motorized vehicles' delays with the difference that bicycles are not likely to queue when approaching intersections due to low volumes of bicycle traffic currently.

Adapting travel forecasting methods from motorized vehicles models for current bicycle models requires the assumption that some value of free flow bicycle speed will be constant along the route and across routes. A "GPS-Based Bicycle Route Choice Model for SanFrancisco, California" assumes an average speed of 10 mph without regard to the route's or the rider's characteristics (Hood, Sall, & Charlton, 2011). The authors report this is weakness of the model due to its exclusion of "dissuasive effects of hills" that are especially a concern in the San Francisco area. Another piece of research that assumes constant speed is "A Bi-Objective Cyclist Route Choice Model," which was conducted in Auckland, New Zealand (Ehrgott, Wang, Raith, & van Houtte, 2012) . The authors assume that speed is constant and that travel time is proportional to distance, but that while topography affects the attractiveness of the route it does not affect travel time.

Several authors have estimated generalized travel speeds for bicyclists (Broach, Gliebe, & Dill, 2011; Forester, 1983; Forester & Forester, 1994; Smith Jr, 1975) . However, they usually contradict each other, basically because of different assumptions on the part of the authors about the rider's level (professional or avocational cyclists), types of bicycle (roadster, commuting bike, etc.), and sample limitations (student athletes only or professional cyclists), etc. A good attempt to study influence of different factors on bicycling speed was performed using a GPS data survey in Minneapolis (El-Geneidy, Krizek, & Iacono, 2007) . The researchers assumed that travel speeds are based on bicycling facility type (regular local street, on-street facility, and off-street facility), personal characteristics (gender, age, ridership comfort level), and segment specific and/or trip characteristics. The variables that were analyzed by the study included: facility type, distance traveled, trip length, segment length, average daily traffic (motor vehicles/day), the number of signalized intersections, morning commute (yes or no), speed, age, gender, and comfort (stress) level. Trip length and segment length have shown significant positive influences on bicycling speed, just as the number of signalized intersections decreases travel speed. An analysis of personal characteristics of riders showed that gender has a significant effect on bicycling speed (men ride 0.67 mph faster than women), and level of comfort also impacts speed (people ride faster when they feel safe), while age has been shown to have a small, if any, effect on travel speed. Although, the study is relevant to the issues explained above, it is limited by very small sample size (8 respondents). It also does not test the effects of ridership frequency and individual physical potential on bicycling speed.

Another interesting study on bicycling speed and acceleration was conducted in Leeds, UK (Parkin & Rotheram, 2010) . The authors completed a GPS study on the relationship between riding speed and road geometry. They collected trip data from sixteen volunteers (including four women). They found that over a range of slopes, from three percent downhill to three percent uphill, the speeds of eighty-five percent of riders varied from 18 kph (about 11 mph) to 25 kph (about 15.5 mph), with speeds on flat terrain averaging 22 kph (about 14 mph). The authors also suggest that downhill slopes do not provide the expected advantage, since people tend to maintain speed at safe levels for road conditions. They also note that the average time of a single trip was in a range between 15 and 50 min.

So, based on previous studies and some intuitive sense as a bicycle rider, one can conclude that riding speed affects the total time of a trip and the topography of a road affects riding speed. However, the intermediate element that lies between speed and topography is the rider. It is the rider who finds riding up steep slope difficult and automatically reduces speed. In fact, different riders will react differently to the same change in topography. As mentioned above, men overcome high hills more easily than women, and people riding for exercise will prefer steeper slopes than commuters do. To explain why this happens I need to introduce human power element.

17

The human body is a complex mechanism. For the purposes of this research, basic definitions and concepts of body operations need to be presented. Bicycle motion is the result of work that being performed by the rider. In physical terms work is "the amount of energy being changed from one form into another by a force" (Franklin, 2010), or simply force applied over the distance traveled. Work is measured in joules (J). Work performed for some amount of time is described by power, and measured in watts (W). The amount of work that can be performed depends on individual's level of energy.

"Energy is the ability to do work" (Faria, 1978). The international unit system (SI) unit of energy is the joule; however, it may also be expressed in kilocalories (kcal). The energy expenditure of human can be divided into two categories: resting energy expenditure (basal fraction), or the amount of energy required to sustain basic body functions, and activity energy expenditure, or the amount of energy used to perform all other functions. Faria argues that "muscle work during cycling is about 25% efficient in converting fuel energy to mechanical work. The remaining energy is dissipated as heat" (p. 38).

Energy expenditure during cycling can also be obtained by measuring oxygen uptake at rest and during cycling. This type of energy refers to aerobic power or aerobic capacity, which is a maximum amount of sustained physiological work that person can do; it is measured by amount of oxygen taken in during exercise. Aerobic power reflects the capacity for a longer exercise period but at lower intensity levels. Faria argues that the "aerobic energy system is the most effective and efficient manner of muscle metabolism." Energy also can be obtained from anaerobic metabolism, a complicated process of burning fuel in human muscles. Anaerobic power is an intense exercise that can be performed for short period of time or a serial sequence of periods that usually last less than 2 minutes. It is important to remember when planning bicycle facilities to try to avoid segments where cyclist is required to apply extreme physical effort to overcome road barriers.

It is obvious that power levels vary for different people based on many characteristics, but that they primarily do due to variances in gender, body mass, and the fitness level of persons. There are many different approaches on how to measure human power (NSCA-National Strength And Conditioning Association).

Two different approaches to measure the energy needed to cycle are found in the literature. David Wilson (Wilson & Papadopoulos, 2004) proposes an equation that allows the determination of power levels and/or speeds of riders based on topography, wind resistance, and rolling resistance. As explained above, power is measured by watts (or joules per second). This formula determines power at a specific point in time. However, using this method we only can measure the maximum speed that person can achieve at a given moment, an approach does not give a tool to measure long-term power or cumulative energy expenditure.

Another approach was proposed by Smith (Smith Jr, 1975) where the physical abilities of a rider were approximated as a fraction of his aerobic work capacity. The method is used to evaluate the acceptability of maximum road grades (on existing roads) for users with different states of physical health. However, the method is hard to be generalized because is based on fraction of aerobic work capacity of the specific person and shows what portion of aerobic power of this person is required to overcome the grade. This approach provides a tool to evaluate long term energy use through the aerobic and anaerobic capacities; however, the method relies on strict assumptions that were criticized by Forester (for example, Smith assumes that riders use three-gear bicycles which are not commonly used by many modern riders). The method developed in early 1970s has not been adopted and used.

Another important characteristic of physical activity is fatigue. Fatigue is developed by an accumulation of lactic acid in muscles and depends on intensity of work. However, light pedaling actually assists recovery as opposed to raising fatigue levels (Faria, 1978), and thus the alternation of cycling at higher and lower intensity levels even for longer distances may result in lower cumulative fatigue than heavy pedaling over short distances. This issue needs to be studied. Moreover, the level of exhaustion is important for safety issues. Research has shown that individuals who endure moderate or greater fatigue experience decrement of balance control and require more cognitive resources to perform attentional tasks (Simoneau, Bégin, & Teasdale, 2006)   .

CONCLUSION

This review has demonstrated that current methods of bicycle infrastructure planning do not account for all factors that are important for cyclists and that affect the decision to cycle and the routes to be chosen. While planners mainly pay attention to safety and the economic feasibility of bicycle projects, there is much more to be integrated into the decision making process. Travel time and ease of cycling are important factors that have to be considered.

However, current bicycling models assume constant speed along a route and across routes, and make the simplifying assumption that travel time is strictly proportional to the distance travelled. This simplistic assumption is not supported by empirical study of riding speed change. GPS studies in the US (El-Geneidy, Krizek, & Iacono, 2007) and UK (Parkin & Rotheram, 2010) have found that riding speeds decrease significantly due to uphill slopes. This change in speed can be explained by the higher level of power required.

This literature review has shown that there is no currently used method that determines realistic cycling travel times based on changes in speed due to topography. However, both time and topography are important elements of cycling, especially for utilitarian cyclists. Introduction of such a method will allow creating facility location strategies with a purpose of minimizing travel times and providing quickest and/or easiest routes between main destinations.

3. RESEARCH QUESTION AND OBJECTIVES

Increasing the number of people who use bicycles as alternative transportation mode is a reasonable purpose for transportation planners. To achieve that purpose, it is important to understand which criteria for the siting of routes and facilities are important for riders and will make them more likely to cycle. This can be done by placing facilities to minimize travel time or physical effort to complete the route since these are important factors that affect the decision to cycle. Being able to realistically forecast travel time for bicyclists will improve the transportation planning process and increase the reliability of bicycling as a transportation mode. It will also allow more efficient placement of bicycle facilities like bike lanes and bike sharing stations. Considering that literature review has shown a lack of current methods for bicycling travel time forecasting the questions of this research are (1) how to estimate realistic travel time, considering the maximum power available to the rider and (2) how to choose bicycle routes based on a realistic minimum travel time for the bicycling trip.

The objectives of this research are:

- Identify the criteria that have the most significant influence on bicycling travel time.
- Develop a model of estimating travel time that accounts for maximum power available to the rider.
- Solve the bicycle route choice problem to minimize travel time or energy needed when more than one route can connect two destinations.
- Propose locations of bikeways based on the quickest or easiest route between destinations.

4. METHODOLOGY

This chapter will explain the methodology used in the research to solve the problem of route choice where travel time is a cost attribute of the route. In this case "route" is a set of contiguous network links connected to two different bikestations, one on each end. The time required to get from one station to another is determined by the time a bicyclist spends riding along the route and the time spent waiting due to delays. In this research travel time on links is the main focus of the analysis, since the delay time is not significantly affected by rider's nature but by traffic regulations. The fastest route in this case is the one that provides the minimum time as a sum of travel time on the links within the route.

$$T_{min} = min_{1 \leq i \leq n}\{T_i\}, \tag{1}$$

where T_{min} = travel time on the fastest route, min
T_i = total travel time on links of route alternative i, min

$$T_i = \sum_{j=1}^{n} t_{ij}, \tag{2}$$

where t_j = travel time on link j of route i;
n = number of links

The method to calculate travel time for bicycles should be significantly different from automobile travel time. In vehicular travel models free flow travel time is calculated based on free flow speed and distance traveled. The free flow speed can be either observed or assumed based on posted speed limits and information about travel behavior. Current transportation models for bicycling use a similar approach, where average bicycling speed is assumed to be constant throughout the route. However, this approach cannot be considered realistic for bicycle transportation. While an automobile's actual speed is constrained more by traffic, signal delays and regulations rather than the vehicle's ability to reach particular speed, bicycling speed is limited to physical abilities of a rider.

The relationship between bicycling speed and human power was studied by (Whitt & Wilson, 1982) and (Wilson & Papadopoulos, 2004) . They suggest that power required from a rider to sustain a particular bicycling speed can be estimated based on physical laws.

Bicycling power is a function of air resistance, rolling resistance, and slope resistance forces and can expressed by an equation (3).

$$W_w = [K_A(V + V_w)^2 + mg(s + C_R)]V, \qquad (3)$$

where W_w = bicycling power, W;

K_A = aerodynamic-drag factor, kg/m;

V = riding velocity, m/s;

V_w = headwind velocity, m/s;

m = mass, calculated as sum of rider's mass and bicycle's mass, kg;

g = acceleration due to gravity, m/s²;

s = slope (rise divided by run);

C_R = coefficient of rolling resistance.

In this case power (W_w) is the power delivered by the driving wheel and is somewhat less than power produced by rider. This difference occurs to transmission inefficiency, however Wilson (2004) suggests that rider power is a reasonable approximation for wheel power; so I will take driving wheel power to be equal to rider power, which will be called bicycling power in this paper.

For the purpose of this research, I will explain each variable. According to (Wilson & Papadopoulos, 2004) , *aerodynamic drag factor* (K_A) depends on the rider's size, riding position, clothing, and air temperature, pressure, and humidity. Although air conditions vary by region, season, or even time of the day, riding position and clothing depends on type of a rider and a bicycle. For an urban utilitarian bicyclist, who rides an upright commuting bike and does not wear tight-fitting clothes, at standard air density at sea level (temperature of 59° F) the aerodynamic drag factor approximately equals 0.3871(Santa Cruz Institute for Particle Physics; Wilson & Papadopoulos, 2004) .

Headwind velocity depends on wind velocity, wind direction and the position of a bicycle according to wind direction. For the purpose of this research, headwind velocity will be excluded, since it is not a constant variable that can be generalized without intensive empirical study for a particular place. However, when applying the model to the specific geographic area, a detailed study on wind direction and velocity can be completed. Equation (3) can be rewritten with regard to the assumption of no headwind velocity.

$$W_w = [K_A V + mg(s + C_R)] \cdot V, \qquad\qquad (3a)$$

The power required to overcome slope resistance is based on the total weight (sum of rider's and bicycle's mass times gravitational acceleration), slope, and the coefficient of rolling resistance. According to Wilson (2004), *coefficient of rolling resistance* depends on tire type and pressure and road surface characteristics. Examples of coefficient of rolling resistance are shown in Table 4.1. For the purpose of this research the coefficient of $C_R = 0.003$ is used considering that a commuting bicycle is being used for the forecast. A different coefficient can be used if another bicycle type is considered to be more likely to be used.

Table 4.1: Characteristics of five types of bicycle and rider. Adopted in part from *Bicycling Science* (Wilson & Papadopoulos, 2004)

	Roadster (Utility bicycle)	Sports bicycle	Road racing bicycle	Commuting HPV	Ultimate HPV
Bicycle mass (kg)	15	11	9	20	15
Rider's mass (kg)	77	75	75	77	75
Rolling resistance coefficient, C_R	0.008	0.004	0.003	0.003	0.002

Rider's mass varies significantly and is easy to determine once there is a specific rider under consideration. The model presented here can be tested for different types of users considering different physical characteristics, including mass. The current model is tested for the average male of 80 kg (176 lbs) and a *bicycle mass* of 15 kg.

Slope data was collected for this research with the use of a geo-information system on a block level. The model uses mean values of slope percent for each segment of road network. This value is not constant through the network and thus plays an important role in power-velocity relationship. When riding up- or downhill bicycling work is done with or against gravity. Riding up steep slopes require significant physical effort, and if the power necessary to sustain speed cannot be produced by the rider, riding speed will drop. Going downhill will result in acceleration without a physical effort from a rider. However, on steep slopes riders tend to start braking once they approach maximum safe speed. This means that downhill slopes would not considerably affect riding speed and thus travel time. Based on this assumption, a *riding velocity* of 22 km/h (14 mph) is a constant value in the model unless this speed cannot be sustained due to significant uphill slope. The value of mean riding velocity is adopted from Parkin and Rotheram (2010). Riding velocity is recalculated in the model when the maximum power required to bicycle exceeds power available to the rider.

The amount of power that a human can generate depends on his/her physical attributes (age, gender, fitness level), the type of exercise, the duration of exercise, and the effort level (maximum, minimum or in between). Researchers have shown that power level tends to decrease significantly after one minute of performance at maximum effort level, and stays somewhat constant between 5 and 60 minutes (Webb, 1964; Whitt & Wilson, 1982) . The example of power distribution is shown in figure 4.1.

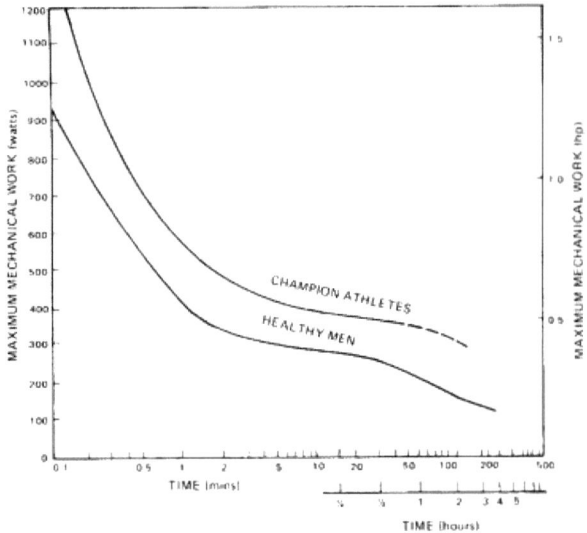

Figure 4.1: Maximum effort of healthy fit men and champion athletes. Reprinted from Webb (1964)

Although the maximum short-term power output of athletes has been studied widely (Faria, 1978; Foster et al., 2003; Hintzy, Belli, Grappe, & and Rouillon, 1999; Macdermid & Stannard, 2012; Morrow, 2005) with the purpose of increasing the performance of professional cyclists, there is not enough data on long-term term power output for different type of cyclists, including people with average athleticism. Whitt and Wilson (1982) suggest that the power output observed for non-athlete cyclists can go as low as 40 W for prolonged periods of time, while data from Webb (1964) shows that "healthy men" can sustain power output of about 300 W over a period shorter than 40 minutes. Parkin & Rotheram (2010) suggest theoretical maximum power output of about 250 W for climbing uphill. The value is calculated based on similar methods to the one presented here, using observed speeds collected by GPS devices, mass, and other characteristics of the riders. The model presented in this paper uses *maximum power output* (W_{max}) of 200 W. This is lower than values observed by Webb (1964) but higher than suggested by Whitt and Wilson (1982) because I assume a value for an average healthy cyclist riding at a comfortable power level. However, I recognize the limitations of this assumption and emphasize that empirical data on power thresholds for different types of users is required for further research.

To summarize, the power model from equation 3a is calculated based on the assumptions explained above, then bicycling power (W_w) is compared to maximum power available to the rider (W_{max}). If $W_w > W_{max}$ riding velocity V is recalculated for the maximum available power ($W_w = W_{max}$). However, the equation for velocity when power is a given is a cubic polynomial and is not easy to use in a route model. Wilson & Papadopoulos (2004) recommend using an iterative approach (see equation 4) with a convergence parameter K_C.

$$V = (\frac{W_{max}}{K_A V^2 + mg(s + C_R)} + K_C V) \cdot \frac{1}{K_C + 1} \tag{4}$$

The final input variables for the power model here are specified in Table 4.2.

Table 4.2: Input variables for Power Model

Variable	Value
Aerodynamic-drag factor (K_A), kg/m	0.3871
Riding velocity (V), m/s	6
Mass (m), kg	95
Acceleration due to gravity (g), m/s^2	9.81
Slope (s)	varies
Coefficient of rolling resistance (C_R)	0.003
Maximum power output (W_{max}), W	200
Convergence parameter (K_C)	0.5

This methodology allows to answer question 2 for the research – how to estimate realistic travel time, considering maximum power available to the rider. The next section of the manuscript will answer question 1 – how to choose a bicycle route based on the realistic minimum travel time of bicycling trip, given available power, by comparing three scenarios for decision making on where to locate bicycle facilities.

5. DATA AND RESEARCH SCENARIOS

The bicycle route choice model in this paper was developed in ArcGIS 10.1 and tested for Washington, DC area. Washington, DC is a known for high rate of bicycle commuting which makes the area attractive for bicycle infrastructure planning. Also, the area has a wide range of elevation changes which allows testing model's assumptions (figure 5.1). The city also has the Capital Bike Share program which features about 175 rental bike stations in the DC region. The particular area for the model run I performed was selected based on a combination of topography and bicycle station locations (figure 5.2). To answer the question addressed by this research, namely how to choose a bicycle route based on a realistic minimum travel time and/or energy expenditure for the bicycling trip, two scenarios of route choice were developed and compared.

Scenario 1: Constant Speed Scenario. A route was selected with the purpose of minimizing time cost where travel time is determined as a function of speed and distance and speed is determined by using a constant value of 16 km/h or 10 mph (this value is adopted from the GPS model for San Francisco Bay area (Hood, Sall, & Charlton, 2011). In this scenario time varies as a linear function based on the distance and the quickest route is also the shortest in terms of distance. The value of total travel time is simply the sum of link travel times considering a riding speed of 16 km/h.

Scenario 2: Power Model Scenario. A route was selected with the purpose of minimizing time cost, but in this model time is a function of speed and distance and speed is a function of power. The value of speed in this scenario is calculated based on the power model explained in previous section of this paper. Because speed drops on uphill slopes, areas with flat or moderate topography are expected to have higher values of speed and shorter travel times, while areas with steep slopes are expected to have lower values for speed. However, travel time in this scenario is not linear in relation to distance or power.

To compare travel times under the two scenarios, ArcGIS Network Analyst extension was used. North America Detailed Streets data from www.arcgis.com was used for street network layer. Elevation data for the DC area was retrieved from www.nationalmap.gov. The locations of bicycle stations were geocoded based on the station map at www.capitalbikeshare.com.

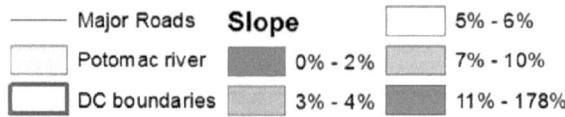

Legend

—— Major Roads	**Slope**	▢ 5% - 6%
▢ Potomac river	■ 0% - 2%	▢ 7% - 10%
▢ DC boundaries	▢ 3% - 4%	■ 11% - 178%

Figure 5.1: Washington, DC area

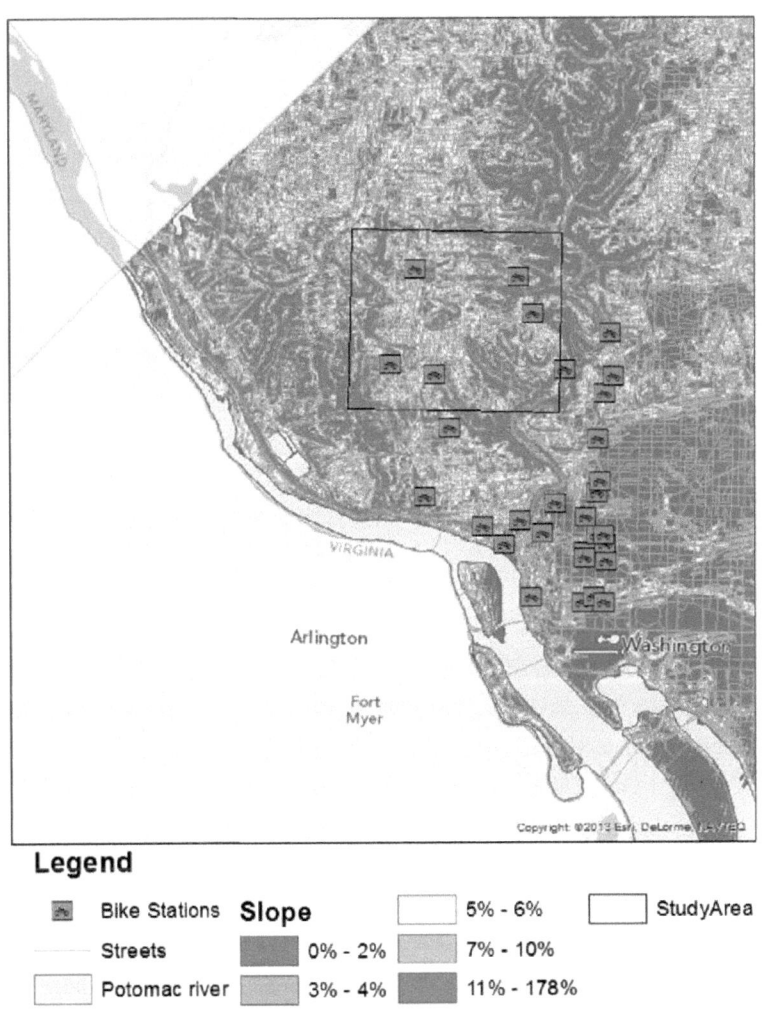

Legend

⊠ Bike Stations	**Slope**		▢ 5% - 6%	▢ StudyArea
— Streets	▧	0% - 2%	▨ 7% - 10%	
▢ Potomac river	▨	3% - 4%	▨ 11% - 178%	

Figure 5.2: Selection of study area

To run the route analysis the original network file was modified and dead-end links were deleted since they do not provide connectivity between destinations. All road types including alleys and driveways were included in the analysis as potential locations for designated bicycle routes. To calculate values of speed and power for the Power Model Scenario the following variables were obtained. The compass direction of each link was assigned for both directions and the slope aspect was calculated to identify uphill slopes. Flat areas or downhill slopes were ignored and speed for these segments was assigned as 22 km/h or 14 mph. Mean average slope for each link was calculated to produce input data for the model. Then the methodology explained in the previous section of this paper was applied to calculate the power output required to maintain constant speed for each link and generate speed values for segments where power needed to maintain maximum safe speed exceeds the maximum value of power that can be produced by the rider. For the Constant Speed Scenario the values of travel time are equal for both direction of the road segment, this why only one value of travel time was calculated for this scenario. However, in the Power Model Scenario the values of speed can be different for two directions if the segment does not have flat topography. This is why two values of speed, one for each direction of the link, were computed. Three values of travel time required to traverse each link were calculated:

1. Travel time as a speed-distance function based on assumption of constant speed of 16 km/h. The value is same for both directions. Further in the paper travel time for the Constant Speed Scenario is called *CS_Time*.

2. Travel time for one direction of the link (From node To node direction) was calculated based on From-To value of speed produced by power model. Below in the paper travel time for the From-To direction of the link calculated based on Power Model Scenario is called *PM_FT_Time*.

3. Travel time for opposite direction of the link (To node From node direction) was calculated based on To-From value of speed produced by power model. Further in the paper travel time for To-From direction of the link calculated based on Power Model Scenario is called *PM_TF_Time*.

Based on values of travel time, I calculated the amount of physical work required from a person to complete one route or another. Power is a work performed over some period of time, so I multiplied power by time and found the work required from the rider in joules. I computed work for time estimates for both directions on each link. Work computed based on the constant speed assumption is further called *CS_Work* and work calculated based on power model is called *PM_Work*.

To compare scenarios I applied Network Analyst extensions of ArcGIS 10.1. Then the quickest routes for each scenario were calculated for every permutation connecting each station of five selected Capital Bike Share stations to every other station for both directions with impedance of *CS_Time* for Scenario 1 and impedance of *PM_Time* for Scenario 2. The Washington, DC road network has many segments with one-way traffic direction restrictions. These restrictions were included into the route choice process and applied to both scenarios. A total of forty routes were compared in pairs for two scenarios.

At the final stage I developed recommendations for bikeway locations. The highest priority was given to the road segments where two or more of the quickest routes are located, road segments that have one of the quickest routes were assigned moderate priority, and links that are not part of any route were given low priority. The detailed discussion of results is given in the next section of this paper.

6. DISCUSSION OF THE RESULTS

This section discusses how route total travel time differs based on the assumption on cycling speed and how this difference affects quickest route selection. For this, I ran two quickest route scenarios where one assumed constant average speed and another accounted for speed difference based on the power model explained above. I then combined quickest route into one network system and proposed priority roads for infrastructure location.

SELECTION OF QUICKEST ROUTE

Routes for both scenarios were compared in pairs to identify differences and similarities. Table 6.1 shows whether two scenarios produced same or different results between pairs of bike stations. The *From* column shows station of origin and *To* row indicate station of destination. When routes for two scenarios completely overlap the matrix indicates "same"; however, when at least some difference along the route occurs the matrix field is assigned to "different".

Table 6.1: Route comparison matrix for two scenarios

To / From	1	2	3	4	5
1	-	same	same	different	different
2	same	-	same	different	different
3	same	different	-	same	same
4	different	different	same	-	same
5	different	same	different	same	-

Examples of routes that are different between the scenario with the constant speed assumption and the scenario with speed based on power model is shown on figure 6.1.

When solving the quickest route problem, Network Analyst searches for the combination of network links that will result into minimum total travel time. When we consider constant speed, in scenario one, travel time is linear based on the shortest distance, which mean that the quickest route equals shortest route between stations. However, if we consider that speed drops when unmanageable physical effort is required from a rider, as in scenario two, then travel time is not linearly related to the distance anymore. In this case a shorter link can actually take more time to pass through than a longer one. Based on the power model, when power required to sustain speed exceeds maximum power available to a rider, cycling speed drops and travel time on this link increases. However, having lower power levels required from rider will not automatically result in a route being quickest. When the road network provides many options for route selection, the route that is easiest in terms of power output may be significantly longer and thus the shorter but more hilly route may be the fastest.

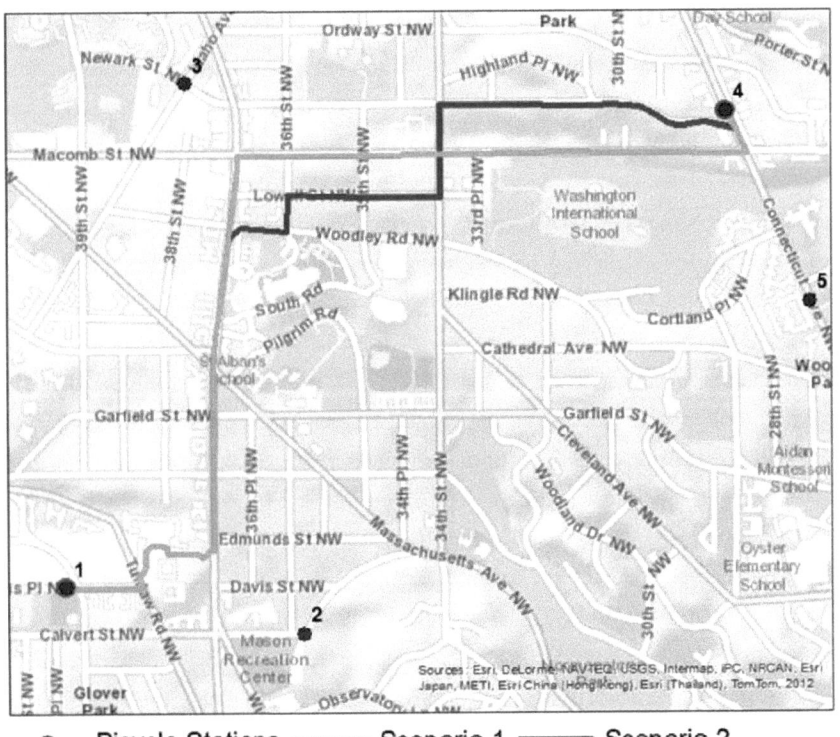

Figure 6.1: Quickest route between stations 1-4 for the two scenarios

To compare the results that the two travel time models produce, I compared travel cost attributes for each route. I calculated total values of all five cost parameters (CS_Time, PM_Time, CS_Work, PM_Work, and Length) for each route, even though only one cost parameter (either *CS_Time* or *PM_Time*) was used to produce the quickest route. Those parameters are: travel time calculated based on constant speed assumption (*CS_Time*), travel time calculated based on power model (*PM_Time*), total length of the route (*Length*), total physical work required to complete the route based on *CS_Time* (*CS_Work*) and total physical work required to complete the route based on *PM_Time* (*PM_Work*).

Table 6.2 shows the cumulative values of the five cost parameters for every pair of routes. Scenario 1 or Scenario 2 identifies the method used to compute the route, where Scenario 1 is the route computed with impedance of travel time based on the constant speed of 10 mph and Scenario 2 is the route computed with the impedance of travel time estimated by the power model. Thus, there are forty routes compared in the table. Values in **bold italics** identify routes between the same pairs of stations which are different for two scenarios. Those routes are of particular interest for this research because they show the differences in two methods of computing travel time.

As expected, the total length of the route in scenario one is always shorter or equal to the total length of the route in scenario two. This is obvious, since in scenario one travel time is proportional to the distance, and in fact the shortest route is being solved. It is also pointless to compare results for travel time, since each route has the minimum time assumed by its model (i.e., a route solved with impedance *CS_Time* will have smaller values of *CS_Time* cost that the route solved with impedance *PM_Time*). However, what is the most interesting is how each route produced by the two different assumptions performs in terms of physical effort required from a person. I compared the difference in physical work required from a rider to under the two scenarios. Values of work are calculated by multiplying power by *CS_Time* and by *PM_Time*. Table 6.3 summarizes results for the routes that are different for the two scenarios. When *PM_Time* attribute is used to calculate work (*PM_Work*) then Scenario One produces significantly higher results for all routes. When *CS_Time* is used to estimate work, then four out of nine routes show lower work values than for scenario one.

Table 6.2: Route comparison for two scenarios

Route	PM Time, min		CS Minutes, min		Length, m		PM Work, J		CS Work, J	
	Scenario 1	Scenario 2	Scenario 1	Scenario 2	Scenario 1	Scenario 2	Scenario 1	Scenario 2	Scenario 1	Scenario 2
1-2	2.43	2.43	2.61	2.61	694.85	694.85	368	368	360	360
1-3	7.4	7.4	7.25	7.25	1932.9	1932.9	1260	1260	1150	1150
1-4	11.1	10.12	11	11.25	2934	3000.7	1820	1460	1654	1482
1-5	10.14	9.71	10.7	10.77	2854.77	2873	1538	1411	1478	1437
2-3	6.86	6.86	6.62	6.62	1765.86	1765.86	1179	1179	1065	1065
2-4	9.28	8.35	8.83	9.03	2353.74	2406.7	1528	1232	1322	1208
2-5	7.93	7.73	8.31	8.34	2216.02	2224.36	1220	1178	1166	1171
3-4	4.57	4.57	5.64	5.64	1505.13	1505.13	575	575	670	670
3-5	5.74	5.74	7.23	7.23	1927.03	1927.03	692	692	830	830
4-5	1.54	1.54	1.98	1.98	529.29	529.29	177	177	220	220
2-1	2.85	2.85	2.61	2.61	694.85	694.85	499	499	423	423
3-1	6.32	6.32	6.66	6.66	1777.24	1777.24	979	979	949	949
3-2	5.44	5.22	6.04	6.24	1610.2	1663.94	820	713	850	805
4-1	11.54	11.47	11	11.51	2934.03	3069.75	1897	1857	1647	1714
4-2	9.59	8.61	8.85	8.91	2360.48	2374.9	1618	1358	1367	1291
4-3	7.08	7.08	5.92	5.92	1579.7	1579.7	1281	1281	1004	1004
5-1	11.67	11.28	10.78	11	2875.93	2932.04	1991	1898	1697	1715
5-2	8.43	8.43	8.39	8.39	2237.19	2237.19	1400	1400	1292	1292
5-3	10.22	9.25	7.81	9.37	2082.7	2497.7	1969	1549	1463	1468
5-4	1.54	1.54	1.98	1.98	529.29	529.29	525	525	377	377

Table 6.3: Cumulative work difference between two scenarios

Route	Scenario 1 - Scenario 2	
	PM_Work, J	CS_Work, J
1-4	360	172
1-5	127	41
2-4	296	114
2-5	42	-5
3-2	107	45
4-1	40	-67
4-2	260	76
5-1	93	-18
5-3	420	-5

On one hand, the assumption tested in Scenario 1 that speed will be average among links simplifies the process of calculating travel time. This assumption may be valid when analyzing existing cycling patterns where the rider does not have universal knowledge about quickest route, or is simply concerned about safety of that route. However, when planning bicycle infrastructure, it is possible to create a system that will allow for significant time savings. Because all routes in this analysis are under twelve minutes, the actual time difference in minutes between travel times computed based on both the power model and the constant speed assumption is under three minutes. With time rising with a linear relationship to distance (scenario one) the direction of travel is not important unless there is restriction of direction (e.g., where there is a one-way street). This might always be true for automobiles, however, two directions of exactly the same route might be very different for a bicycle rider due to topography. Figure 6.2 shows that there is no significant difference of geographical location between route 3-5 and 5-3 under Scenario 1; Minutes = 7.23 and Minutes = 7.81 respectively. However, when speed varies based on power and topography the results differ significantly. Total travel time based on the power model varies almost by 5 minutes between two directions (Time = 5.74 and Time = 10.22) and total work performed by rider goes up from 692 J (route 3-5) to 1969 J (route 5-3).

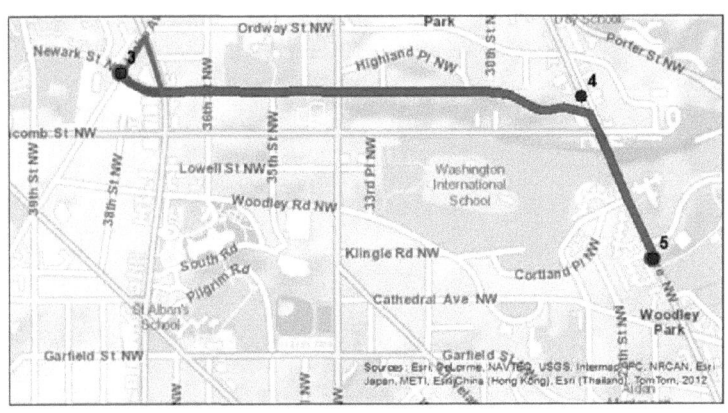

Legend

●   Bicycle Stations   ——— Route 5-3   ▬▬▬ Route 3-5

Figure 6.2: Scenario 1 Route Comparison

VALIDATION OF THE RESULTS

My methods infer that the power model produces more precise calculations of travel time. To test whether my results are accurate, I had a rider perform test rides for two of the actual routes in my model. The routes 3-5 and 5-3, discussed above, were picked to validate my results. A twenty-nine year old male physically fit regular but not professional bicyclist completed a set of three rides for each route and total travel time for each ride was recorded by a mobile application, www.endomondo.com. The application allows stop recording travel time every time when a bicyclist makes a stop. Thanks to this, riding conditions were set as close to model assumptions as possible. Figures 6.3 and 6.4 show the routes that rider accomplished (exactly the same as produced by Scenario 2) and table 6.4 provides information on travel time for each ride.

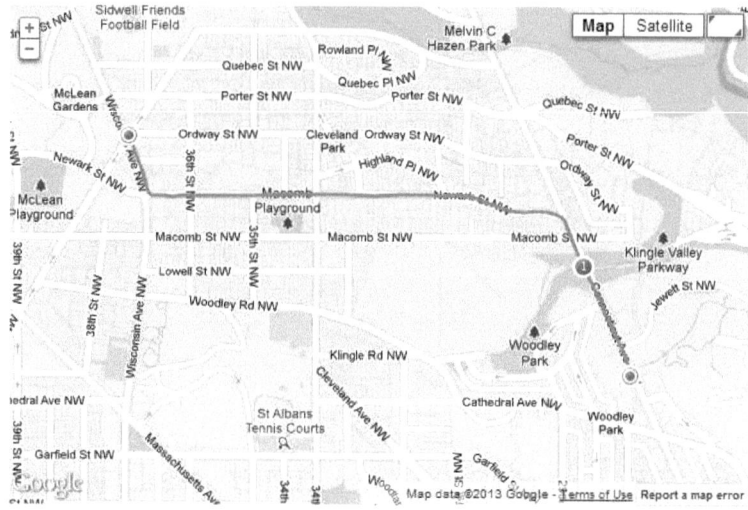

Figure 6.3: Validation of travel time results, Route 3-5

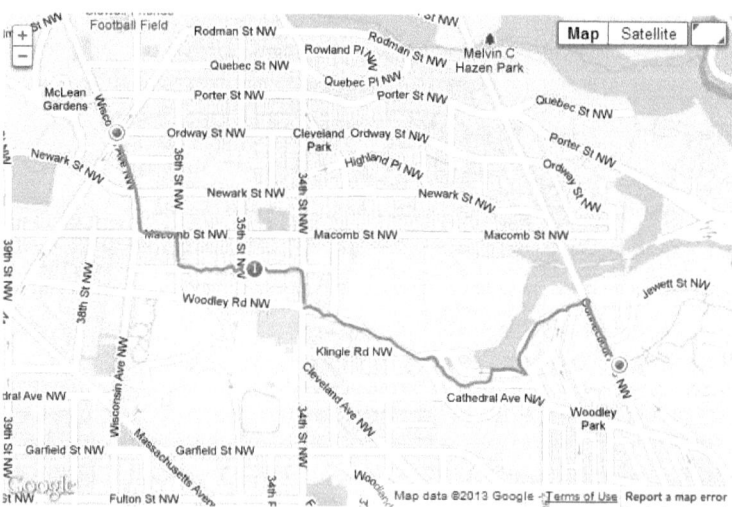

Figure 6.4: Validation of travel time results, Route 5-3

39

Table 6.4: Validation of travel time results

Route 3-5	Route 5-3
6 min 1 sec	9 min 52 sec
6 min 23 sec	9 min 43 sec
6 min 15 sec	10 min 14 sec

To compare, the value for travel time calculated by the power model for route 3-5 equals 5 min 45 sec and for route 5-3 it equals 9 min 15 sec. Higher values for the theoretical as opposed to the observed time for route 5-3 can be explained by cumulative fatigue experienced by the rider. The rider reported that he got tired by the end of the experiment. Otherwise, test rides show that results produced by the theoretical model are adequate and can be used for further analysis.

From there, I used travel time computed based on the power model to develop recommendations for bicycle infrastructure prioritization. I also used *Power x Time* impedance to produce a route network that minimizes total physical work required from a rider to complete the route.

METHODS OF INFRASTRUCTURE PRIORITIZATION

I combined all the quickest routes calculated based on the power model for five stations of capital bikeshare program into one infrastructure map. Road segments that accommodate two or more quickest routes are given high priority, segments where one quickest route is located are given moderate priority, and streets that are not part of any route are given low priority.

However, the quickest routes do not necessarily go through the same road segments for both directions. This means that if one side of the road is part of one or more of the quickest routes, and the other side of the road is not, then there is no need to develop infrastructure on both sides of the road. This why I split the bikeway prioritization maps into two maps based on direction. In figure 6.5, the From-To direction shows combination of ways for routes 1-2, 1-3, 1-4, 1-5, 2-3, 2-4, 2-5, 3-4, 3-5, and 4-5. This map indicates segments where routes go on the right side of the road. Figure 6.6, which shows the To-From direction, assembles routes 5-4, 5-3, 5-2, 5-1, 4-3, 4-2, 4-1, 3-2, 3-1, and 2-1.

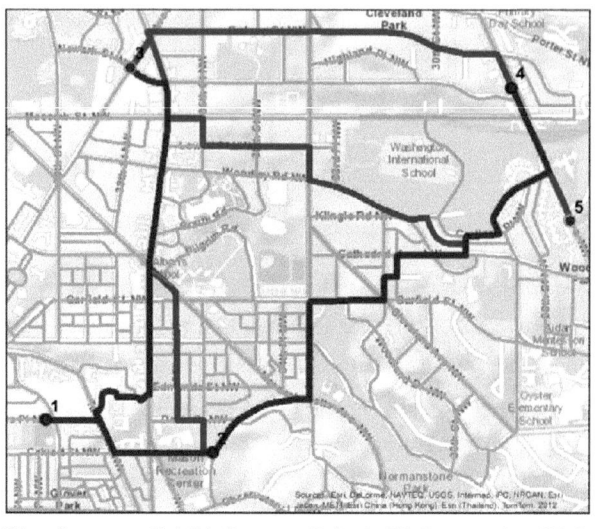

Figure 6.5: Bikeway prioritization based on quickest route method (From-To direction).

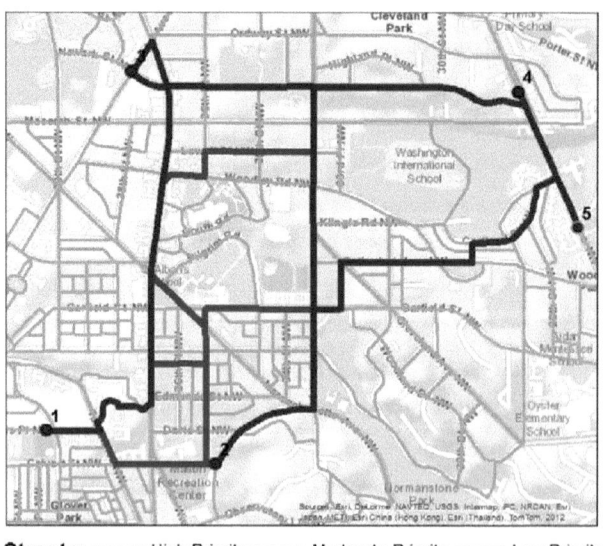

Figure 6.6: Bikeway prioritization based on quickest route method (To-From direction).

This paper also provides an alternative route network based on minimization of physical work (Figure 6.7 – 6.8). This network was developed by the same method as quickest route system but instead of *PM_Time*, I used the *PM_Work* impedance. This network focuses on providing access through the least challenging routes within a network. Both the quickest and easiest route methods account for power and are highly related to each other. However, the results are not identical since time required to complete the route is being minimized in one case and in total work required to accomplish the route being minimized in another.

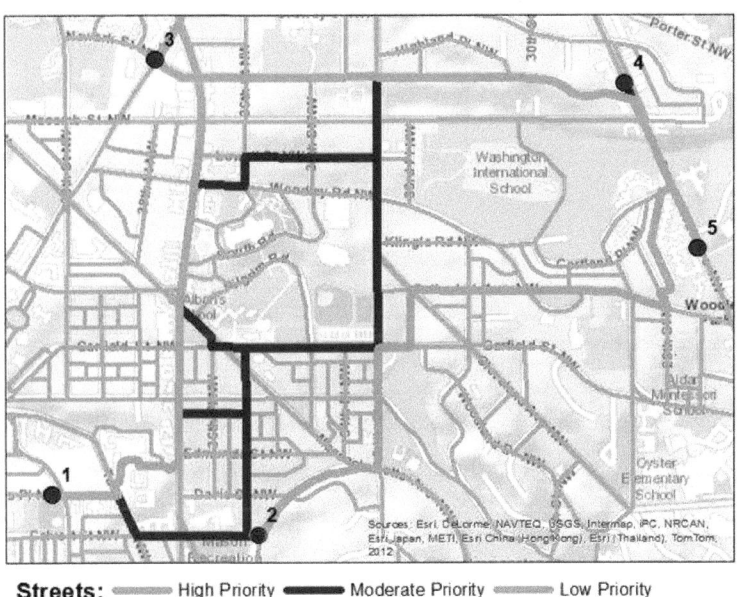

Streets: ━━ High Priority ━━ Moderate Priority ━━ Low Priority

Figure 6.7: Bikeway prioritization based on easiest route method (From-To direction).

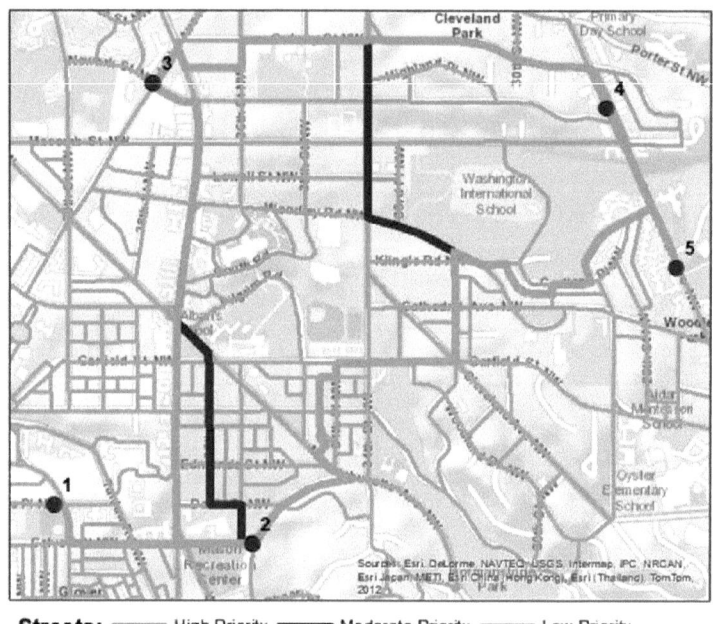

Figure 6.8: Bikeway prioritization based on easiest route method (To-From direction).

Although time impedance was used to build quickest route network and work impedance was used to build easiest route network, the results are not significantly different. The quickest route sometimes involves more direct routes than the easiest route. However, it is up to planning staff to decide what method to use in their particular situation.

CONCLUSION

Development of bicycle infrastructure will remain a relevant goal for many cities in the US for several decades. Lack of funding requires transportation planners to come up with new methods to prioritize placement of infrastructure to reduce cost and increase benefits of the bicycle network utilization. The literature review above has shown that safety, travel time, and ease of cycling are important factors that affect people's decisions to bicycle and, thus, shape bicycle travel patterns. Safety of bicycling was intensively studied in recent decades and intelligent tools for evaluating and planning for safety were developed. However, current models make simplistic assumptions about bicycling speeds that lead to rough values of travel time that are not always true.

The research here offers a method that allows one to predict realistic travel time based on change in speed due to topography. The model presented in the paper allows estimating bicycling speed based on relationship between human power and road topography. It then makes it possible to calculate value of travel time and solve the quickest route problem between key destinations. Values of power required to achieve or sustain particular speeds, were then converted into physical work that needs to be produced to complete the route. Finally I offer a method on how to prioritize location of bicycle road facilities based on either quickest or easiest routes of the network.

APPLICATION TO PRACTICE

This research is a small piece in a large process of planning bicycle transportation. The most important findings of this research are that it allows the calculation of more precise values of bicycling travel time that can be used in travel forecasting models, travel planning for bikeshare systems, as a time estimate tool for integrating transit and cycling, as a planning tool for prioritization of bikeway infrastructure, etc. Another finding arising from this method is the possibility to calculate cumulative work or relative difficulty of the route, which can be used by communities that try to implement bicycling as an active living tool. After minor modifications this model can be used to evaluate the levels of difficulty for different routes and to promote recreational ways for users with different levels of physical health. The model can also be integrated into personalized route planning software. However, some limitations of this study have to be addressed.

The first limitation that I faced while developing power model is a lack of good empirical data on aerobic (long-term) power for those who are not professional athletes. This research provides a valid average value to estimate travel time for an average rider. However, empirical study of power levels at different comfort levels for different groups of people has to be accomplished to bring my model to the next level of sophistication.

Also, the travel time model does not account for delays at intersections. Including this data into the model may significantly change route allocation if there are intersections with long durations for red signals and/or significant numbers of intersections. However, it is expected that travel time in both scenarios will be affected equally by delay function and thus it does not affect credibility of the method presented in this paper.

Last but not least, I did not have enough time and resources to complete the validation of the model with a large enough sample of time tests. However, test rides for two routes with a sample of three rides for each route has shown results close to the model results.

NEXT STEPS

There are many ways this research can go. However, I see a necessity of addressing the limitations listed above before moving forward. After a signal delay function is integrated into the travel time model, the final version of quickest route choice can be produced. I then want to integrate other factors important to people such as directness of the route, aesthetics, safety, etc. and to develop a bicycle accessibility model.

REFERENCES

American Association of State Highway and Transportation Officials. (2012). *Guide for the development of bicycle facilities* (4th Edition ed.). Washington, DC: American Association of State Highway & Transportation Officials. Retrieved from from Pascal

Application to track bike ride. Retrieved, 2013, from www.endomondo.com

Broach, J., Gliebe, J., & Dill, J. (2011). Bicycle route choice model developed using revealed preference GPS data. Paper presented at the *90th Annual Meeting of the Transportation Research Board, Washington, DC,* Retrieved from ftp://ftp.hsrc.unc.edu/pub/TRB2011/data/papers/11-3901.pdf

Ehrgott, M., Wang, J. Y. T., Raith, A., & van Houtte, C. (2012). A bi-objective cyclist route choice model. *Transportation Research Part A: Policy and Practice, 46*(4), 652-663. Retrieved from http://www.cmnzl.co.nz/assets/sm/4740/61/1330D-Raith.pdf

El-Geneidy, A. M., Krizek, K. J., & Iacono, M. J. (2007). Predicting bicycle travel speeds along different facilities using GPS data: A proof of concept model. Paper presented at the *Proceedings of the 86th Annual Meeting of the Transportation Research Board, Compendium of Papers,*

Faria, I. E. (1978). *Cycling physiology for the serious cyclist* Thomas.

Forester, J. (1983). *Bicycle transportation*. Cambridge, Mass.: MIT Press.

Forester, J., & Forester, J. (1994). *Bicycle transportation: A handbook for cycling transportation engineers* (2nd ed.). Cambridge, Mass.: MIT Press.

Foster, C., de Koning, J. J., Hettinga, F., Lampen, J., La Clair, K. L., Dodge, C., . . . Porcari, J. P. (2003). Pattern of energy expenditure during simulated competition. *Medicine and Science in Sports and Exercise, 35*(5), 826-831. Retrieved from http://www.uwlax.edu/faculty/foster/fosterweb/Foster%20%20Pattern%20of%20Energy%20Expenditure.pdf

Franklin, K. (2010). *Introduction to biological physics for the health and life sciences*. Chichester, West Sussex: Wiley.

Fried, B. (2009, 3/11). LaHood to bike advocates: U.S. DOT will be your "Full partner". *DC.STREETS.BLOG.ORG*. Retrieved from http://dc.streetsblog.org/2009/03/11/lahood-to-bike-advocates-us-dot-will-be-your-full-partner/

Harkey, D., Reinfurt, D., Knuiman, M., Stewart, J., & Sorton, A. (1998). *Development of the bicycle compatibility index: A level of service concept.* (Final Report No. FHWA-RD-98-072).U. S. Department of Transportation.

Hintzy, F., Belli, A., Grappe, F., & and Rouillon, J. (1999). Optimal pedalling velocity characteristics during maximal and submaximal cycling in humans. *EUROPEAN JOURNAL OF APPLIED PHYSIOLOGY AND OCCUPATIONAL PHYSIOLOGY, Volume 79*(Number 5 (1999)), 426-432.

Hood, J., Sall, E., & Charlton, B. (2011). A GPS-based bicycle route choice model for san francisco, california. *Transportation Letters: The International Journal of Transportation Research, 3*(1), 63-75.

Jaffe, E. (2012, What's the best way to figure out what bike riders really want?

King, M. (2002). Bicycle facility selection: A comparison of approaches. Retrieved from http://nacto.org/wp-content/uploads/2011/03/Bicycle-Facility-Selection-A-Comparison-of-Approaches-2002.pdf

Litman, T., Blair, R., Demopoulos, W., Eddy, N., Fritzel, A., Laidlaw, D., Forster, K. (2005). *Pedestrian and bicycle planning: A guide to best practices* Victoria Transport Policy Institute.

Macdermid, P. W., & Stannard, S. (2012). Mechanical work and physiological responses to simulated cross country mountain bike racing.

Morrow, J. R. (2005). *Measurement and evaluation in human performance* (3rd ed.). Champaign, IL: Human Kinetics. Retrieved from http://www.loc.gov/catdir/toc/ecip0422/2004020218.html

National Association of City Transportation Officials. (2012). NACTO urban bikeway design guide. Retrieved 3/1 from http://nacto.org/cities-for-cycling/design-guide/

Nettler, J. (2013, 3/3). U.S. DOT to develop its own bike and pedestrian safety standards. *Planetizen.* Retrieved from http://www.planetizen.com/node/61010

NSCA-National Strength And Conditioning Association.*NSCA's guide to tests and assessments* Human Kinetics.

Parkin, J., & Rotheram, J. (2010). Design speeds and acceleration characteristics of bicycle traffic for use in planning, design and appraisal. *Transport Policy, 17*(5), 335-341. doi: 10.1016/j.tranpol.2010.03.001

Pucher, J., Dill, J., & Handy, S. (2010). Infrastructure, programs, and policies to increase bicycling: An international review. *Preventive Medicine, 50*, S106-S125. Retrieved from http://nacto.org/wp-content/uploads/2011/03/Infrastructure-Programs-and-Policies-to-Increase-Bicycling-An-International-Review.pdf

Santa Cruz Institute for Particle Physics. Retrieved 2/18, 2013, from http://scipp.ucsc.edu/outreach/balloon/atmos/1976%20Standard%20Atmosphere.htm

Sener, I. N., Eluru, N., & Bhat, C. R. (2009a). An analysis of bicycle route choice preferences in texas, US. *Transportation, 36*(5), 511-539.

Sener, I. N., Eluru, N., & Bhat, C. R. (2009b). An analysis of bicyclists and bicycling characteristics: Who, why, and how much are they bicycling. Paper presented at the *Transport Research Board Annual Meeting. Washington DC.*

Simoneau, M., Bégin, F., & Teasdale, N. (2006). The effects of moderate fatigue on dynamic balance control and attentional demands. *Journal of Neuroengineering and Rehabilitation, 3*(1), 22. Retrieved from http://www.jneuroengrehab.com/content/3/1/22

Smith Jr, D. T. (1975). *Safety and locational criteria for bicycle facilities. user manual volume I: Bicycle facility location criteria.* (No. FHWA-RD-75-113).Department of Transportation. Federal Highway Administration.

Stinson, M. A., & Bhat, C. R. (2003). Commuter bicyclist route choice: Analysis using a stated preference survey. *Transportation Research Record: Journal of the Transportation Research Board, 1828*(-1), 107-115.

Taylor, D., & Davis, W. J. (1999). Review of basic research in bicycle traffic science, traffic operations, and facility design. *Transportation Research Record: Journal of the Transportation Research Board, 1674*(-1), 102-110.

Tolley, R. (2003). *Sustainable transport: Planning for walking and cycling in urban environments* CRC.

U.S. Federal Highway Administration, American Traffic Safety Services Association, American Association of State Highway and Transportation Officials, & Institute of Transportation Engineers. (2010). *Manual on uniform traffic control devices for streets and highways (MUTCD)* (2009th ed.). Washington, D.C.: U.S. Dept. of Transportation, Federal Highway Administration. Retrieved from http://mutcd.fhwa.dot.gov/pdfs/2009/mutcd2009edition.pdf

US Department of Transportation. (2010). *The national bicycling and walking study: 15–Year status report.* (). Washington, DC: US Department of Transportation Federal Administration. Retrieved from http://katana.hsrc.unc.edu/cms/downloads/15-year_report.pdf

US Department of Transportation. (2012). *Traffic safety facts 2010.* (). Washington, DC: US Department of Transportation National Highway Traffic Administration.

Webb, P. (1964). *Bioastronautics data book* Scientific and Technical Information Division, National Aeronautics and Space Administration.

Whitt, F. R., & Wilson, D. G. (1982). *Bicycling science* (No. Monograph)

Wilson, D. G., & Papadopoulos, J. (2004). *Bicycling science* The MIT Press.

Winters, M. L. (2011). *Improving public health through active transportation: Understanding the influence of the built environment on decisions to travel by bicycle.* University of British Columbia). Retrieved from https://circle.ubc.ca/bitstream/handle/2429/33377/ubc_2011_spring_winters_meghan.pdf?sequence=5

Zheng, F. (2011). *Modelling urban travel times. thesis.* Delft University of Technology). (TRAIL Thesis Series T2011/9)

Printed by Books on Demand GmbH, Norderstedt / Germany